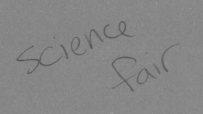

Science
fair

Party Science

Peter Pentland and Pennie Stoyles

CHELSEA HOUSE
PUBLISHERS

A Haights Cross Communications Company
Philadelphia

This edition first published in 2003 in the United States of America by Chelsea House Publishers, a subsidiary of Haights Cross Communications.

Chelsea House Publishers
1974 Sproul Road, Suite 400
Broomall, PA 19008-0914

The Chelsea House world wide web address is www.chelseahouse.com

Library of Congress Cataloging-in-Publication Data

Pentland, Peter.
  Party science / by Peter Pentland and Pennie Stoyles.
  p. cm. — (Science and scientists)

  Includes index.
  Summary: Describes the scientific principles behind various objects and activities at a party including balloons, fireworks, magic, and laughter.

  ISBN 0-7910-7015-8
  1. Science—Experiments—Juvenile literature. [1. Science—Miscellanea.] I. Stoyles, Pennie. II. Title.
  Q164 .P3685 2003
  507.8—dc21

                                                    2002001281

First published in 2002 by
MACMILLAN EDUCATION AUSTRALIA PTY LTD
627 Chapel Street, South Yarra, Australia, 3141

Copyright © Peter Pentland and Pennie Stoyles 2002
Copyright in photographs © individual photographers as credited

Edited by Sally Woollett
Text design by Nina Sanadze
Cover design by Nina Sanadze
Illustrations by Pat Kermode, Purple Rabbit Productions

Printed in China

Acknowledgements
Cover: Fireworks, courtesy of Steve Lovegrove.

Australian Picture Library/Bettmann/Reuter, p. 29; Coo-ee Picture Library, pp. 5 (top right), 6–7, 9 (top), 11 (right), 12–13 (bottom); Getty Images/Photodisc, pp. 4–5 (bottom), 5 (bottom right), 8, 9 (bottom), 10–11 (center), 18–19, 20–21; Imageaddict, pp. 10 (left), 24–25; Steve Lovegrove, pp. 13 (top), 16–17, 22, 28; NASA Glenn Research Center, p. 17 (center and right); Bob Barker Magician/Newspix, pp. 26–27; Dale Mann/Retrospect, pp. 7 (right), 14–15, 23.

While every care has been taken to trace and acknowledge copyright the publisher tenders their apologies for any accidental infringement where copyright has proved untraceable.

Contents

Glossary words

When a word is printed in bold you can look up its meaning in the Glossary on page 31.

Science terms

When a word appears like this **dissolved** you can find out more about it in the science term box located nearby.

Have you ever wondered...

...what makes the fizz in soft drinks?

...how fireworks get their colors?

...what makes you laugh?

...how magic tricks work?

Did you know that all the answers have something to do with science?

Science at your party

Parties are all about fun. People, families and communities have parties to celebrate special occasions like birthdays, anniversaries and festivals. At parties you have special foods and drinks. You make special decorations and you sometimes have entertainment like fireworks, light displays, clowns or magicians.

There is a lot of science going on at a party. Soft drinks fizz because they contain gases, which are under pressure until you take the lid off. Ice cream and Jell-O have interesting chemical structures, which give them their particular texture and taste. When fireworks go off, explosive **chemical reactions** happen. Similar explosions happen when you make popcorn.

Scientists

Food scientists develop fun party foods like ice cream, soft drinks and sweets. Soft drink bottles are made of plastics developed by chemists to withstand the pressure of the 'fizz'. Computer scientists and explosives experts work together to make beautiful fireworks displays. Party music and lighting are possible because of the work of electricians and electrical engineers. Scientists who find out about the human brain can explain how magicians trick people, why people laugh, and why people sometimes vomit.

There are many types of scientists and they all have different jobs to do.

- Electrical engineers design lighting and sound systems.
- Gelotologists study laughter.
- Pyrotechnicians study fireworks.

In this book you will:

- look at the science of party foods such as ice cream, Jell-O, popcorn and soft drinks

- discover how candles, sparklers and party lights work

- meet someone who makes science fun

- find out what happens inside your body when you laugh or sometimes vomit.

Where do soft drink bubbles come from?

At a party you often have fizzy soft drinks. Have you ever wondered why soft drinks only fizz when you take the lid off? Where does the fizz come from?

Bubbles of gas

Soft drinks are made from water and sugar. When you put sugar in water, it mixes in so well that you cannot see the sugar any more. The sugar has **dissolved** and the mixture of sugar and water is called a **solution**. Colors and flavors are added to the solution.

The fizz comes from carbon dioxide gas, which is pumped in. Carbon dioxide gas dissolves in the drink, a bit like the way the sugar dissolves. You cannot see the carbon dioxide gas when it is dissolved. So when you buy soft drink you cannot see any bubbles of gas in it. When you take the lid off, there is a hissing noise and the bubbles of carbon dioxide gas 'undissolve' and bubble out of the solution.

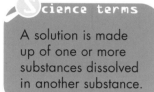

Science terms

A solution is made up of one or more substances dissolved in another substance.

Science fact

Fizzy facts

• A liter of soft drink contains about 25 teaspoons of sugar.

• Soda water is just water and carbon dioxide without any flavor or color added.

• Tonic water contains quinine, which is an antipyretic. An antipyretic is any substance that reduces fever.

The fizz from a soft drink explodes in the air.

Under pressure

The carbon dioxide gas stays in soft drink only when you keep the lid of the bottle tightly closed. The gas wants to escape but it cannot because there is nowhere for it to go. When you open the lid, the pressure is released and the gas can escape.

Soft drink bottles and lids must be able to stand up to high pressure. Most soft drink bottles are made from either glass or a plastic called polyethylene terephthalate (or PET for short). The very first soft drink bottles were made of glass and had cork stoppers. Cork shrinks if it dries out, so the bottles had to lie on their sides. The soft drink would always be touching the cork and it would stay damp and not shrink. Some companies even made the soft drink bottles with round bottoms so you had to lie them down.

Carbon dioxide

Carbon dioxide is a colorless gas that is heavier than air. All animals, including humans, breathe out carbon dioxide as a waste product, but plants need it to grow. You can make carbon dioxide at home by mixing vinegar with baking soda. The mixture fizzes because a **chemical reaction** happens and the bubbles produced are carbon dioxide gas.

Science term

A chemical reaction is a change in one or more substances to form new substances.

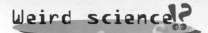

Try this 👍

Dancing raisins are a good party trick.

1 Put some soda water in a glass.

2 Drop in a handful of raisins and watch what happens. At first the raisins will sink to the bottom, but eventually bubbles will stick to them and they will float to the surface. At the surface, the bubbles fall off, so the raisins sink. Then they rise and sink again.

3 See how long the raisins dance up and down in the soda.

'Dancing' raisins rise and fall with the gas in the soda water.

Party food that pops and fizzes

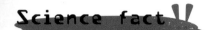
The trick to perfect popcorn

Have you ever noticed that when you make popcorn, some of the corn seeds do not pop? All seed coatings have a tiny hole in them. The hole is there to let water into the seed when you put it in the ground. If you heat up corn seeds slowly, the water turns into steam slowly. The steam can escape through the tiny hole, and it never builds up enough pressure to pop the corn. So the trick to making popcorn is to heat the seeds quickly.

Popcorn pressure

When you make popcorn, you are making a lot of little explosions happen. The science behind this is that steam pressure is being made. Popcorn is made from the dried seeds of corn cobs. Most (but not all) of the water is removed from them. The corn seeds have a tough outer coating. When you make popcorn you heat the corn seeds up quickly. The tiny bit of water that is left in the seed heats up quickly and turns into steam. The steam builds up pressure inside the corn seed. Eventually the pressure gets so high that the seed coat breaks and the corn seed explodes.

Fizzy Powder

Pop Rocks are a patented type of candy. They contain many of the ingredients found in hard candies: sugar, corn syrup, water, and flavoring. But Pop Rocks have a special ingredient—carbon dioxide. To 'gasify' the candy, ingredients are heated under high pressure to about 280 degrees Fahrenheit (137 degrees Celsius). Then carbon dioxide gas is added. The syrup is cooled quickly while still under pressure, trapping tiny carbon dioxide bubbles inside the sugar mixture. Once the candy is cooled, the pressure is released. The candy shatters, but the small pieces still contain bubbles of carbon dioxide.

When you put Pop Rocks on your tongue, the heat of your mouth starts to melt the sugar. The carbon dioxide bubbles can then explode through the thinner candy walls. You hear and feel these explosions in your mouth.

The pressure of steam turns corn seeds into popcorn.

Have you ever had an ice cream cake that has been baked in the oven? This is a special party cake that is hot and cold at the same time. Bombe Alaska has cake on the bottom and ice cream on the top, which is then completely covered in meringue. Meringue is made of eggwhites and sugar. This mixture has been beaten until it is really fluffy. Thousands of tiny air bubbles get trapped in the eggwhites.

Bombe Alaska is a cake with its own insulation.

The cake is then baked quickly in a hot oven to brown the meringue. The trick with Bombe Alaska is to get the cake and ice cream really cold before you start. When you put it in the oven, the cake on the bottom and the meringue on the top and sides protect the ice cream from the heat of the oven. The tiny air bubbles in the cake and the meringue are good insulators. The cake and the meringue stop the heat being moved to the inside of the cake and melting the ice cream.

Insulators

Anything that contains trapped air is a very good heat insulator. Blankets and quilts keep you warm because they contain trapped air. Polystyrene is so light because it contains tiny trapped bubbles of air. It is used to make cups that keep hot drinks hot.

Whisking air into eggwhites creates a good insulator.

How cool is ice cream?

Ice cream is a good party food. Ice cream is frozen foam. Homemade ice cream is usually made of cream, eggs, sugar and flavors (such as fruit). You stir, whip or beat the mixture to trap tiny air bubbles that make the foam. Then you freeze it.

Ice cream before refrigerators

Nobody really knows when ice cream was first invented, but people were making ice cream a long time before refrigerators existed. So how did they get it to freeze?

Water **freezes** and forms ice at 32 degrees Fahrenheit (0 degrees Celsius). When you add salt to ice, the mixture melts and becomes much colder (down to –26 degrees Fahrenheit or –32 degrees Celsius). Before refrigerators, people used to get ice delivered to their homes. They froze the ice cream ingredients by stirring them in a container cooled with a mixture of salt and ice.

To make ice cream you can sit a bowl of cream, sugar and eggs on top of another bowl full of ice and salty water. You keep whisking (stirring) the ice cream mixture and the salty ice will eventually freeze the ice cream solid.

Science term

The freezing point of a substance is the temperature at which it turns from a liquid into a solid.

Science fact

Ice cream scoop

The first ice cream cones were introduced at a world fair in St Louis in 1904. A waffle maker and an ice cream seller were working in booths next to each other. They had the idea to roll the waffles into a cone shape and put in a scoop of ice cream so that people could walk around and eat the ice cream with one hand free.

Science fact

Freezing temperatures

One of the uses of negative (or 'minus') numbers is to describe temperatures that are colder than 0 degrees. Minus numbers are shown by putting a minus sign (–) in front of the number.

Ice cream contains trapped air bubbles.

What makes ice cream taste good?

Ice cream is usually made from milk or cream, which contain different amounts of fat. The more fat, the smoother and richer the ice cream. Sometimes powdered milk is added to help the ice cream hold air when it is whipped. Sugar makes the ice cream sweet and brings out the fruit flavors. Adding sugar also makes the ice cream softer. Air bubbles in the ice cream make it fluffier and feel less cold in your mouth. Ice cream can contain up to 50 percent air.

The next time you eat an ice cream, look at the ingredients on the package. Ice cream that you buy in the store can contain many strange-sounding things such as stabilizers and **emulsifiers**. Food scientists add stabilizers to make ice cream firmer to chew. If stabilizers were not added, big ice crystals would form in the ice cream and it would not taste smooth and creamy. Food scientists also add emulsifiers to stop the fat part of the ice cream from separating from the water part. In homemade ice cream, eggs are the emulsifiers. In commercial ice cream the emulsifiers used are chemicals such as monoglycerides, diglycerides and polysorbates.

Weird science!?

Two of the stabilizers used in ice cream are carrageenan, which is **extracted** from red seaweed, and sodium alginate, which is extracted from another seaweed, brown kelp.

Science term

An emulsion is a mixture of oil and water that does not separate out. This is because it contains an emulsifier.

A substance in brown kelp is used to make ice cream.

Gelatin: wobbly, foamy and chewy

Jell-O is a popular brand of gelatin dessert. There are hundreds of different dessert recipes that include Jell-O. Some people even eat Jell-O as a side dish with a meal or use it in a salad recipe. Mousses are like whipped-up Jell-O. Jelly beans, marshmallows and other candies are made like really stiff Jell-O. All of these wobbly, foamy and chewy desserts have a common ingredient: gelatin.

What is gelatin?

Gelatin is a type of **protein**. It is a liquid when it is warm, but it sets into a sort of wobbly solid when it cools down. This is because of the shape of the gelatin particles, or molecules. They are quite long and tangled, like chains.

How gelatin works to make Jell-O

Do you know how to make Jell-O? A package of Jell-O is a fine powder. The powder is a mixture of gelatin, sugar, flavoring and coloring. Gelatin gives Jell-O its wobbliness. You dissolve the Jell-O powder in hot water and then leave the liquid to cool. When the gelatin dissolves in hot water, the long gelatin chains untangle and float around in the liquid. When you put this liquid into the refrigerator, the long chains start to link together into a tangled web. The Jell-O becomes a sort of solid and can be molded into different shapes.

Jell-O is not a solid solid, it is a wobbly solid. If it is a warm day, the Jell-O melts easily, because the links between the long chains are weak and are easily broken.

How else does gelatin work?

Chocolate mousse

Chocolate mousse is a fluffy, foamy sort of dessert. It is made from whipped-up cream and chocolate. Gelatin is added to make the dessert set so that the air bubbles stay trapped in the foam. Ice cream is also a foam, but freezing is used instead of gelatin to trap the air bubbles.

Jelly beans, gumdrops and other candies

Have you ever toasted a marshmallow or eaten a gummi worm? Many different candies contain gelatin to make them chewy.

Gumdrops, jelly beans and gummi worms contain more gelatin than Jell-O does. More gelatin means more long chains tangled together, making these candies more chewy and less wobbly than Jell-O.

Marshmallows are made from whipped-up eggwhite, sugar, gelatin and water. Marshmallows get their name from a plant called a marsh mallow. The roots of this plant contain a jelly-like substance. It used to be used instead of gelatin to make these candies. One of the oldest known sweets is Turkish delight (or Lukumi). It is made of sugar and gelatin and is flavored with lemon or rosewater and sometimes contains pistachio nuts.

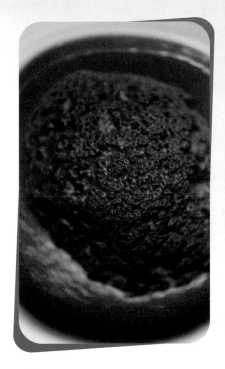

The gelatin in chocolate mousse traps air bubbles.

Thickening and coating

Other foods that contain gelatin include cream cheese, soups and gravies, canned meats and thickened cream. Gelatin is also used to make coatings for some pills. The coating makes the pills easy to swallow. When the pill reaches your stomach, the gelatin dissolves and the medicine inside the pill is released.

As a gelatin dessert cools, it takes on the shape of its container.

Balloons

Have you ever blown up a balloon so big that it burst? Do you know why helium balloons float up into the air if you let them go? Did you know that you can put a skewer through a balloon without bursting it?

How balloons are made

Balloons are made from a liquid rubber called latex, which is very stretchy. Latex comes from the sap of rubber trees that grow in tropical areas. Latex looks like milk when it comes out of the trees. It is mixed with water, dyes and some other chemicals before the balloons are made. In the balloon factory, there are lots of balloon molds. They are dipped into the latex mixture, which sticks to the molds. They are then passed through revolving brushes which make the rims on the top of the balloons. Finally they are dried in an oven, to make the latex solid, and then the balloons are peeled off the molds.

Why does rubber stretch?

Rubber stretches because of the shape of the rubber molecules. They are very long and tangled. Imagine a whole bowl of cooked spaghetti (without sauce) and imagine that each strand is linked to other strands at certain points along the strand. This is what rubber would look like under a very powerful microscope. You could stretch it a certain amount and the strands would not break, but if you stretched it too much the links between the long strands would break and the balloon would burst.

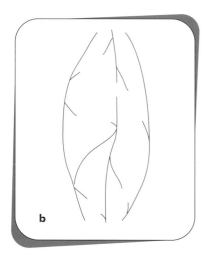

(a) The rubber strands in a blown-up balloon are closely linked together. (b) A balloon bursts because the strands of rubber break.

Balloons are made by dipping molds into a latex mixture.

14

Helium-filled balloons

When you blow up a balloon with your breath, you fill it with air from your lungs. The gas that goes into the balloon is mainly carbon dioxide. Carbon dioxide will fill the balloon, but it will not make it float. But if you fill the balloon with a gas that is lighter than air, such as helium, the balloon will float up.

Helium is very light because helium particles are tiny. Helium is found underground with natural gas. It Is separated out and put into gas bottles. The helium particles are so small that they can escape through the tangle of the balloon's rubber molecules more easily than carbon dioxide. The process is a bit like putting peas in with spaghetti; the peas work their way through the tangle and end up at the bottom of the bowl. This is why helium-filled balloons go flat more quickly. The helium particles have escaped.

Mylar is a sort of plastic that looks silvery like foil. It is sometimes used for helium-filled party balloons. These balloons float for much longer than helium-filled rubber balloons because the helium cannot escape as easily through the Mylar.

Try this

Have you ever popped a balloon with a pin? It makes a big bang. If you are very careful, you can put a bamboo skewer through a blown-up balloon without bursting it. There is a trick to how you do it.

1 Do not blow the balloon up too far. Tie a knot in the end.

2 Coat the bamboo skewer thinly with oil or vaseline.

3 Start at the knot end of the balloon and gently twist the skewer into the rubber. Gently push it right through and twist it out near the other end where the balloon rubber is thickest. (You may need a few tries to get it right.)

If you are very careful, the skewer pushes the long stretchy rubber molecules out of the way, without breaking the links between them. If you imagine the spaghetti again, you can imagine gently putting a knife right into a bowl of spaghetti without cutting any of the strands because they slide out of the way.

On your birthday you have candles on your birthday cake. You can also use candles to decorate your house for a party or your table for a dinner party. At Halloween people sometimes put a candle inside a carved out pumpkin to make a scary face. Candles are also used at many religious celebrations.

How do candles work?

There are two parts to a candle: the wick and the wax. Wax is a fat that is solid at room temperature and melts at about 120 to 140 degrees Fahrenheit (50 to 60 degrees Celsius). Like other fats, wax burns when it is lit. In the middle of the candle there is a piece of string or cloth called the wick. Lighting the wick heats up the nearby wax and turns it into a liquid. This small pool of melted wax moves up the wick. When it reaches the flame, the liquid wax turns into wax **vapor**. It is the wax vapor that burns. When you blow out a candle, you see a stream of white 'smoke', which is wax vapor.

hot air rises

flame stays vertical

convection currents

Convection currents make the candle flame rise up.

You can see the pool of melted wax at the base of the wick of this candle.

Famous scientist

Michael Faraday

In 1826 an English scientist named Michael Faraday (1791–1867) started giving lectures for children. In 1860 he gave six talks called 'The Chemical History of the Candle'. The Christmas lectures have been running ever since.

Faraday wrote a book called *The Chemical History of the Candle*, which was published in 1860. Apart from being famous for making close observations of candles, he also discovered the link between magnetism and electricity.

Candle flames and air currents

Have you ever noticed that no matter how you hold a candle the flame always goes straight up? **Convection** currents are what make candle flames go up. The rising hot air drags the flame up into a teardrop shape.

Whenever you heat a gas or a liquid, the surrounding air moves about in convection currents. The candle heats the air, which rises up. Eventually it rises up so far away from the candle that it cools down and is pushed away by the hot air behind it. The cooling air then falls down again. You can make candle decorations that make use of convection currents. Their propellers twirl around because of the currents that are caused by the hot air rising.

Trick birthday candles

Blowing out all your birthday candles in one breath is supposed to be lucky. You can buy candles that seem to blow out and then re-light themselves. These candles have a special wick, which contains a chemical called magnesium. When you blow out the candle, there is enough heat left in the wick to start the magnesium burning and the candle seems to re-light itself. The secret to the trick is that the candle really did not get put out properly in the first place.

Science term

Convection is the movement of heat from one place to another.

Science fact

Space candle

What do you think would happen if you lit a candle in space? In the weightlessness of space, convection currents do not exist. The candle burns with a spherical (ball-shaped) flame.

The shape of a candle's flame on Earth (left) is different to its shape in space (right).

Many people know the sign of the zodiac for their birthday and they read their **horoscopes**. Each of the signs of the zodiac represents a pattern of stars in the sky. There are many more of these groups of stars in the sky, and they are called constellations.

Patterns in the sky

Over thousands of years, people have looked up into the night sky and 'joined the dots' to make imaginary pictures in the sky. They have also made up stories and myths to go with the pictures.

Astronomy and astrology

Ancient astronomers believed that the Sun, Moon and planets were symbols of the gods. The planets were given the names of Greek and Roman gods. These heavenly bodies were believed to influence the lives of kings and the fortunes of their countries. Ancient Chinese emperors believed that the heavens sent signs and good omens for their rule. The ancient Greeks thought that the stars and planets influenced everybody, not just kings and emperors.

In the late 1500s the famous astronomer Tycho Brahe made his living by making predictions, called horoscopes, for rich and famous people. At the same time he made measurements of the movements of the stars and planets. These days people who make predictions and write horoscopes are called astrologers. Scientists who measure and study planets, stars and other objects in space are called astronomers.

This is the constellation of Scorpio.

The stars in the sky mean different things to different people.

Constellations then and now

Originally the word *constellation* was the name used to describe a pattern of stars in the sky, for example a cross or a scorpion. Since 1930, scientists have thought about the sky as if it were divided up into 88 different sections, a bit like a big uneven patchwork quilt. Each of these sections is now known as a constellation. Each constellation is named after the biggest group of stars it contains, but it also contains all the other stars and galaxies nearby. The Southern Cross is the smallest of all the 88 constellations.

Signs of the zodiac

The Sun appears to move across the sky on an imaginary path called the ecliptic. If you draw this imaginary path over the top of the stars, the Sun appears to pass through 13 constellations. These are called the zodiac constellations. On the day you were born, the Sun was in one of the 13 constellations, and that is your star sign.

Most astrologers work out star signs and horoscopes using star charts that were drawn more than 2,000 years ago. At that time the Sun only appeared to pass through 12 constellations. This is because the planets and stars in the universe are always moving, and so our solar system has changed position in the universe over time. So the Sun now appears on a different background of stars than it did when the star charts were first drawn.

Because the Sun now appears to pass through 13 constellations instead of 12, the dates of the zodiac have changed.

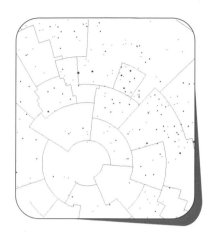

Star maps show where the constellations are in relation to each other.

Signs and dates of the zodiac

Astrological zodiac	Sign	Astronomical zodiac
March 21 – April 19	Aries	April 22 – May 16
April 20 – May 20	Taurus	May 17 – June 20
May 21 – June 20	Gemini	June 21 – July 17
June 21 – July 22	Cancer	July 18 – August 6
July 23 – August 22	Leo	August 7 – September 14
August 23 – September 22	Virgo	September 15 – November 1
September 23 – October 22	Libra	November 2 – November 23
October 23 – November 21	Scorpio	November 24 – November 30
	Ophiuchus	December 1 – December 16
November 22 – December 21	Sagittarius	December 17 – January 16
December 22 – January 19	Capricorn	January 17 – February 13
January 20 – February 18	Aquarius	February 14 – March 9
February 19 – March 20	Pisces	March 10 – April 21

Party lights

When you have a party or celebration, sometimes you decorate your house with strings of colored lights. Light is a type of **energy**. It is the only type of energy that human eyes are sensitive to.

Christmas tree lights

Early Christmas trees were decorated with very thin candles that were wound around or clipped on to the ends of the branches. This was quite dangerous because the candle flames were believed to have started many house fires.

How a lightbulb works

The lightbulbs in party lights are called incandescent bulbs. When you plug the lights in, electricity flows through a tiny coil of wire, called a filament, inside the glass bulb. The wire is made of tungsten metal. Tungsten has a high resistance. This means that the electricity has to work very hard to get through the wire. It works so hard that the wire heats up and glows. Some of the electrical energy is changed into light energy.

Bright ideas

In 1882, a man named Edward Johnson made the first electric Christmas lights. Johnson was a business partner of Thomas Edison. Thomas Edison invented the lightbulb in 1879.

When bulbs are made, the air inside them is sucked out and replaced with a mixture of nitrogen and argon gas. This stops the metal filament from burning up, which would happen if there was oxygen inside the bulb. Eventually, after you have used the bulb for many hours, the metal in the filament will break. Then the electricity cannot flow anymore.

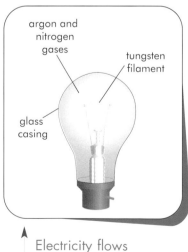

argon and nitrogen gases

tungsten filament

glass casing

Electricity flows through the filament of a lightbulb.

Party lights contain incandescent bulbs.

Electric circuits

Have you ever taken out your party lights and found that they did not work? You have to check every bulb because one faulty bulb means that none of the lights works.

All electrical appliances (including bulbs) use current electricity. For current electricity to work, you need two things. First, you need a circuit. An electric circuit is an unbroken pathway that will allow electricity to flow through it. Secondly, you need a **force** to 'drive' the electricity. The force 'pushes' the electricity around the circuit. Batteries and power points are used to provide the 'push'.

Party lights are usually in a type of circuit called a series circuit. In a series circuit the bulbs are connected one after the other and there is only one pathway for the electricity to flow through. In each bulb electrical energy is turned into light energy and heat energy. If one of the lightbulbs has a broken filament, or if it is not screwed into the socket properly, then the circuit is broken and the electricity cannot flow until the bulb is screwed in or replaced.

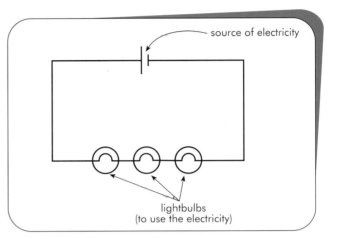

source of electricity

lightbulbs
(to use the electricity)

Party lights are connected in a pathway called a series circuit.

Fireworks

Many people love watching fireworks because of the beautiful colors and patterns made as the fireworks explode. Other people love the loud noise of the explosions. You can also smell the fireworks after they have exploded. People who design and display fireworks are called pyrotechnicians.

Sound and light

If you are watching a fireworks display from a distance, you will probably see the colored light before you hear the bang of the explosion. This is because light travels through the air to you much faster than sound. Sound and light are both forms of energy. One of the big rules in science is that you cannot create or destroy energy, you can just turn it from one type of energy into another. So how are sound and light produced from a firework?

Explosions

Fireworks are explosions. Explosions are a type of chemical reaction. Chemical energy is turned into sound and light energy. When the solid chemicals in the fireworks react together, huge amounts of gases are produced. If the explosion reaction happens in a closed container, the gas pressure builds up and up until the container blows apart with great force, causing a huge bang.

Chemical reactions cause fireworks to explode and chemicals give fireworks their colors.

How do fireworks get their color?

Pyrotechnicians use powdered metals like aluminium, iron and magnesium to make the gold and silver lights. The colors are made by adding different salts. The salt used in food is a chemical called sodium chloride. When sodium chloride burns it makes a yellow flame. Other salts with sodium in them are added to fireworks to make a yellow explosion. Salts with copper in them make fireworks blue. Strontium salts are used to make red fireworks, and barium salts make them green.

Meet a science communicator

Have you ever asked yourself what it would be like to work in a museum? What sorts of jobs are there? How is it possible to get a job? What do you have to study in high school and at college?

Meet Vera Gin

Vera Gin has the answers to these questions. Parties are all about having fun, and you have already seen how much science there is at parties. Vera Gin has run science birthday parties for children. She is now a Visitor Programs Officer at a science musuem, and her job is to help people to see the fun in science.

Vera Gin.

Vera and her family migrated from South Africa to Australia when Vera was 12 years old. She went to a local high school in Melbourne. When Vera left school she studied science at a university and decided to become a science communicator.

A science communicator helps to make scientific ideas easy to understand and fun to learn. At a university science center Vera presented fun science shows for school children. She also worked at the Melbourne Aquarium, where she talked to school groups and other visitors and ran aquarium birthday parties for children.

Vera now works at a science museum. She writes and runs science shows and works out fun science activities for visitors to do. She also helps to develop new exhibitions. Part of her time is spent with visitors and part of her time is spent at her desk organizing and planning.

Vera loves her job as a science communicator. She likes explaining science in a way that is easy to understand. She likes seeing people enjoying themselves so much that they do not even know that they are learning science!

When you have a party, you often have music in the background or for dancing. When you celebrate you sometimes sing special songs, depending on the occasion. Sometimes you have party whistles and party poppers to add to your party noise.

What is sound?

Sounds are important because people use them to communicate with each other. Sounds can make you feel sad, calm, happy or excited. That is why sound is used at parties.

Sound is a type of energy. It is a wave of air that is caused by something that is moved or **vibrated**. Sounds are vibrations. Very fast vibrations make high-pitched sounds and slower vibrations make low-pitched sounds. Your ears pick up the vibrations traveling through the air. The vibrating air goes into your ear and hits your eardrum. This vibrates which, in turn, sets off a chain reaction of vibrations through your ear bones to the inner ear. The vibrations are turned into nerve signals, which are sent to your brain. Your brain understands the signals as sounds.

Decibels

In science, loudness is measured in decibels. Scientists called sound engineers measure and study sound. The symbol for decibels is dB. The decibel scale starts at zero which is the softest sound a human can hear. An increase of one decibel means that the sound has 10 times more energy. A sound of 130 dB is painful to human ears.

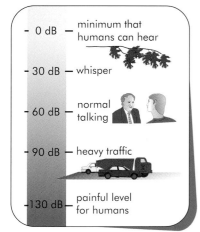

Some sounds are louder than others.

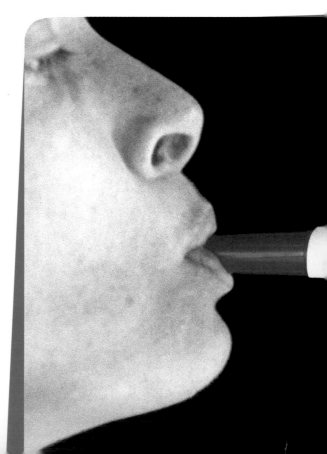

When you blow into a party whistle, you are causing vibrations in the air.

Making music

Party whistles work like all musical instruments. To make a sound you must make something vibrate. Usually it is part of the instrument. In a guitar, you make the string vibrate. In a drum it is the skin, and in a clarinet there is a reed in the mouthpiece that vibrates. Party whistles work a bit like clarinets. When you blow air through them, part of the mouthpiece vibrates.

As well as producing a vibration, musical instruments can **amplify** sounds by increasing the amount of vibration. This is done through the shape and size of the instrument. The sound box in a guitar or violin, the body of a drum and the tube of a trumpet or clarinet increase the amount of vibration, which makes the sound louder.

How a speaker makes sound

A speaker changes electrical energy into sound. The speaker has a cone that is made of a light material that can vibrate easily, such as paper or plastic. It hangs in a metal frame called the basket. The cone is attached to the basket by a flexible rim, so that it can move. At the bottom of the cone there is an electromagnet and a permanent magnet.

An electromagnet is only magnetic when electricity flows through it. When this happens the electromagnet is attracted to the permanent magnet for an instant and the cone moves back and forth. Lots of rapid pulses of electricity make the cone vibrate, and sounds are produced.

Science fact

Musical body parts

Trumpets make sound in a different way. The instrument itself does not vibrate, it is your lips that vibrate against the mouthpiece.

When you sing, your voice produces sound like a musical instrument. Air from your lungs passes over your vocal cords, which vibrate. You change the shape of your mouth to amplify the sound.

Speakers change electrical energy into sound.

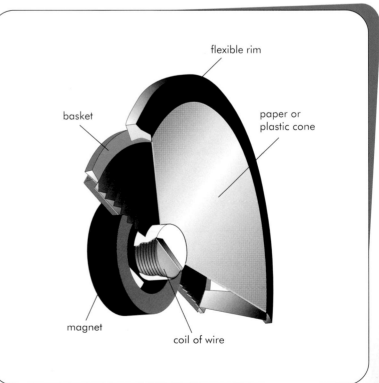

flexible rim

basket

paper or plastic cone

magnet

coil of wire

25

Party tricks and illusions

Have you ever wondered how magic tricks work? Magicians are very careful not to give their secrets away. Magic tricks are entertaining because magicians seem to have special powers and they appear to do things that are impossible, such as sawing someone in half or turning a scarf into a rabbit. Tricks work because the magician creates illusions. Magicians make you think one thing is happening, when really they are doing something else.

Science term

Perception to know what is going on around you involves using your five senses.

Perception

To understand how many illusions happen, you first need to know about **perception**. Your senses send information to your brain and your brain processes the information using your past experience. Past experience allows your brain to take shortcuts through the enormous amounts of information it gets all the time. Many tricks and illusions work because they take advantage of these 'thinking shortcuts'.

Try this 👍

Sometimes your eyes get clashing messages. Your brain gets confused and creates an optical illusion.

1 Roll a sheet of paper up into a tube.

2 Hold the tube up to one eye.

3 Hold your other hand right next to the tube with your palm facing you.

4 With both eyes open, look through the tube and at your hand. Your brain is getting two different messages, which it combines to make the illusion of a hole in your hand!

You can trick your own eyes to see a hole in your hand.

What you do.

What you see.

a hole in your hand!

Tricking your eyes

Most magicians' tricks rely on tricking your sense of sight. This is because humans rely on their sense of sight most. You get so much information from your eyes that you usually trust your sight over your other senses, such as hearing, touch, taste and smell.

Your eyes can be fooled when a movement is too fast, hidden or new. When you see a person getting into a box and being 'sawn in half', you assume that the saw is sharp and the box is a certain shape. Your brain processes what is happening based on what you already know about boxes and saws.

Mathematical tricks

Some tricks are based on mathematics. The person who is doing the trick seems to have the power to read your mind, but they are just relying on counting and simple mathematics.

Try this

1 Ask a friend to think of a number without telling you what it is.

2 Then ask them to double it.

3 Then ask them to add on a number that you choose, say 10 (it must be an even number).

4 Then ask them to divide their answer by two.

5 The last step is to ask them to take away the number they first thought of.

6 Then you tell them the answer. In this case the answer is five.

No matter what the number was that they first thought of, the answer to this trick will always be half the number you chose for them to add on. (Half of 10 is 5.)

Magicians are good at creating illusions.

Laughter

Weird science!?

Tickling makes you laugh. However, no matter how hard you try, you cannot tickle yourself. For tickling to work, you must feel a mixture of pleasure and anxiety. The pleasure comes from the touch of the person tickling you and the anxiety comes from not knowing where you are going to be tickled next.

Laughter is good for you.

What happens when you laugh? A laugh begins with a deep breath in. You then breathe out in 'bits and pieces'. As the air rushes past your vocal cords, the laughing sound is produced. Some people laugh through their mouths and others laugh through their noses. If you laugh a lot, your face goes red. Laughter seems to be catching. If one person starts laughing, other people laugh too.

What is laughter?

Scientists know a lot about laughter, but there is also a lot that they do not know. A person who studies laughter is called a gelotologist. Gelotologists have found out that different parts of the brain control laughter. One part of your brain 'gets' jokes, another part makes you feel happy. Other parts of your brain make your face muscles move and make your heartbeat and breathing become faster. When you laugh, your blood flows faster, which is why your face sometimes goes red. Your tear ducts work harder than usual and the extra tears make your eyes sparkle. Sometimes you laugh so much that you cry.

Laughter is the best medicine

Laughter can make you less stressed. It can also make you produce more of the chemicals that fight disease (antibodies). Laughter gives your face and stomach muscles a good workout. It also helps you to release negative emotions so you feel better. Some hospitals have realized that laughter is very good for your health so they have 'clown doctors' to help to speed up patients' recovery.

Too much fun

Have you ever had so much fun that you were sick? Sometimes when you run around too much, play spinning games or eat too much party food you vomit. Most people think that vomiting is caused by your stomach, but it is really caused by your brain.

Not everyone enjoys rollercoaster rides.

Vomit center

In a part of your brain is your vomit center. When it goes into action, you vomit. It gets its information from different parts of your body. Your stomach sends information to the vomit center when you eat too much at a party or when you have eaten food that is bad or poisonous. If you are on a rollercoaster ride, sailing or even driving, you sometimes feel sick. This is because the balance part of your inner ear alerts the vomit center. Other parts of your brain can send messages to the vomit center. If you have a bad headache or head injury you might be sick. If you see unpleasant sights or smell nasty smells you can vomit. You might even feel sick just reading this section.

Party science timeline

This timeline shows some important party science events. See if you can imagine some of the things that might happen in party science in the future.

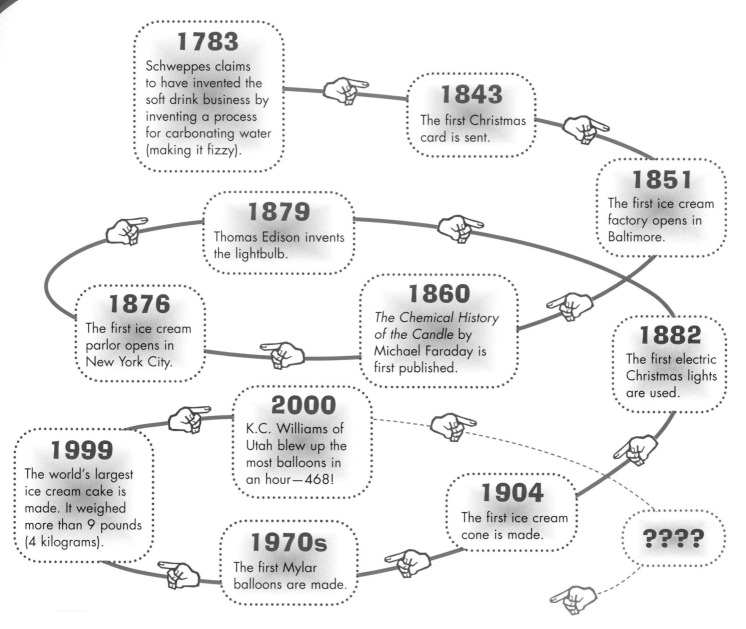

1783
Schweppes claims to have invented the soft drink business by inventing a process for carbonating water (making it fizzy).

1843
The first Christmas card is sent.

1851
The first ice cream factory opens in Baltimore.

1879
Thomas Edison invents the lightbulb.

1876
The first ice cream parlor opens in New York City.

1860
The Chemical History of the Candle by Michael Faraday is first published.

1882
The first electric Christmas lights are used.

2000
K.C. Williams of Utah blew up the most balloons in an hour—468!

1999
The world's largest ice cream cake is made. It weighed more than 9 pounds (4 kilograms).

1970s
The first Mylar balloons are made.

1904
The first ice cream cone is made.

????

What are scientists working on now?

- Scientists are trying to make drinking straws made of starch, which can be eaten. If the straws are accidentally dropped, they can be eaten by animals or can decay to make a natural fertilizer.

- Scientists are working on an exciting new kind of rollercoaster ride. The ride will be able to move in many different directions.

Glossary

amplify	increase volume or strength
by-product	a second product that is made when you are making something else
chemical reactions	changes in one or more substances to form new substances
convection	the movement of heat from one place to another
dissolved	mixed into a liquid so that it seems to disappear
energy	the ability of an object to do work. Energy cannot be created or destroyed, but it can be changed from one form to another
extracted	removed or taken out
force	a push or pull. It can change the movement of an object
horoscopes	forecasts of your future based on where the stars and planets were when you were born
protein	the name for a group of chemicals found in all living things
solution	one or more substances dissolved in another substance
vapor	tiny particles suspended in the air
vibrated	shaken from side to side

Index